MW01515574

EX BIBLIOTECA
D. A. de VLLOA

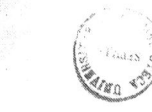

A

BENGAL ATLAS:

CONTAINING

M A P S

OF THE

THEATRE OF WAR AND COMMERCE

ON THAT SIDE OF

HINDOOSTAN.

Compiled from the Original Surveys; and publiſhed by Order of the Honorable the Court of Directors for the Affairs of the East India Company,

By JAMES RENNELL,

Late Major of Engineers, and Surveyor General in Bengal.

M DCC LXXX.

ADVERTISEMENT.

THE Intent of publiſhing the Maps of BENGAL, &c. under the preſent Form, was to render them portable to thoſe who travel over that extenſive Country. A large Map is exceedingly incommodious either in a Tent, Budgerow, or Palankeen: and if divided, without a proper Regard being had to the natural Diviſions of the Country, much Confuſion is likely to enſue for Want of a clear Idea of the relative Poſitions of the ſeveral Parts. I have endeavoured to avoid this Evil, by taking for the Boundaries of my Diviſions, either ſome noted River, Road, or Chain of Hills; without regarding the diſproportionate Sizes, or irregular Figures of the Tracts contained in the ſeveral Maps. In the Lower Part of BENGAL, where a Multitude of Rivers and Creeks interſect the Country in almoſt every Direction, I have caſt the two Diviſions with a View to render the Geography of the inland Navigation as diſtinct as poſſible. Accordingly, the firſt Map comprizes the whole *ſouthern* Navigation; or that between CALCUTTA, MOORSHEDABAD, DACCA, LUCKIPOUR, CHITTIGONG, and TIPERAH reſpectively: the other contains the *midland* and *eaſtern* Navigation; or that between MOORSHEDABAD, DACCA, MAULDAH, SEEBGUNGE, SILHET, and TIPERAH. Each of the other Six Diviſions contains ſuch a Tract, as by its Situation and natural Boundaries, will comprehend the probable Extent of the Seat of War * in that particular Part of the Country.

The BENGAL Provinces lying on the Weſt of the Calcutta and Coſſimbazar Rivers, and South of the Ganges, compoſe Two of theſe Diviſions; of which the one lies on the North, the other on the South of the River Adji.

* The Scale of theſe Maps is confeſſedly much too narrow for Military Purpoſes; but muſt ſuffice until the Provincial Maps can be engraved: and as theſe cannot be contained in leſs than Eighteen Sheets of Imperial Paper, the Time of their Publication muſt neceſſarily be very remote.

The

The northern Provinces of BENGAL compoſe another Diviſion, which includes alſo the BOOTAN Frontier.

The two Diviſions of BAHAR formed by the Courſe of the Ganges, have each a ſeparate Map; that on the North includes the Hills of MOCAUMPOUR and MORUNG: and that on the South extends to the Hills of PALAMOW and RAMGUR; which were conſidered as our Boundary, previous to the Reduction of thoſe Provinces by Major Camac.

PALAMOW, RAMGUR, CHUTA-NAGPOUR, and their Dependencies, are comprized in one Map, which makes the eighth and laſt Diviſion of BENGAL and BAHAR.

The Order of placing the Diviſions being quite arbitrary, I have begun with that which contains CALCUTTA, and gone on towards PATNA; it being the Route of all others, the moſt uſed. The Index Map will at once convey an Idea of the relative Poſitions of the ſeveral Diviſions, as well as their Places in the Book.

As it was neceſſary to bring all BENGAL and BAHAR into one View, I have alſo conſtructed a General Map of thoſe Countries on a more confined Scale, though large enough to contain every Place of the leaſt Note.

The Countries ſituated between BENGAL and DELHI form a ſecond general Map on a ſimilar Scale with that of BENGAL: but it is in few Reſpects ſo compleat; the Survey being conducted on a more limited Plan. Probably theſe Maps contain a larger Tract of ſurveyed Country, than is to be found in all the Maps of the European Kingdoms put together; and they owe their Exiſtence chiefly to the Arrangements made by the late noble Lord, to whoſe Genius and Courage Great Britain owes the Sovereignty of Bengal.

The DOO-AB, COSSIMBUZAR ISLAND, and the Environs of DACCA, have each a ſeparate Map on a large Scale, as being Tracts more particularly intereſting.

AREA of the BENGAL PROVINCES,

IN SQUARE BRITISH MILES.

N. B. A ſquare Mile contains 640 Acres: or 1936 Beagers of 1600 ſquare Yards each.

B E N G A L.

		Square Miles.
Ahmirabads		127
Attyah		787
Birboom		3,858
Biſſunpour		1,256
Burbazzoo		468
Burdwan		5,174
Caugmahry		374
Chittigong		2,987
Chogong		51
Chunacally		269
Chundli		180
Coos-Beyhar		1,302
Dacca (proper)	13,567	
Curribarry	869	
Sundeep	167	
Bominy	56	15,397
Deccan Shabazpour	337	
Hattiah	165	
Iſlands in the Mouth of the Ganges	236	
Dinagepour	3,289	
Callygong	70	3,519
Bajoohow	160	
Duttya-Janguirpour		33
Futtaſing		259
Goragot		1,232
Hoogly and Injellee	1,798	
Company's Lands	882	2,818
Saatſi	138	
Janguirpour	203	
Barbuckpour	159	597
Moſeedah	153	
Pooſtole	82	
Jeſſore		1,365
Iſlamabad		62
Kiſhenagur		3,151

		Square Miles.
Luſkerpour		499
Mahmudſhi		844
Mauldah		168
Midnapour		6,102
Pachete		2,779
Purneah	4,978	
Boherrah	39	5,119
Delawrpour	102	
Purruah		24
Rajemal	2,042	
Coſſimpour	42	2,217
Malduar	56	
Surore	77	
Rangamatty		2,629
Raujeſhy proper	4,071	
Bettooriah	3,942	
Booſnah	2,230	
Pookareeah	711	
Baharbund	520	
Bittrebund	221	12,909
Patladah	487	
Surroopour	249	
Cotwally-hoſſain-pour	65	
Barbuckſing	81	
Shahjole	331	
Rungpour	2,161	2,679
Bootiſhazary	518	
Shilberis		264
Silhet		2,861
Sunderbunds		6,183
Tarpour		83
Tiperah low Lands	1,368	6,618
Woods	5,250	
	Total of Bengal	97,244

BAHAR.

		Square Miles.
Bahar (proper) -		6,680
Bettyah - - -		2,546
Chuta-Nagpour	6,965	9,329
Burwah -	552	
Toree - -	1,022	
Koondah -	790	
Hajypour - -		2,782
Monghir, viz.		
Boglipour -	2,817	8,270
Curruckpour and Hendooah	2,696	
Curruckdea, &c.	2,757	

	Square Miles.
Palamow - -	4,137
Ramgur - - -	5,087
Rotas - - -	3,680
Sarun - - -	2,560
Shawabad - -	1,869
Tyroot - - -	5,033
Total of Bahar	51,973
Total of Bengal and Bahar	149,217

CON-

C O N T E N T S.

No. I. The Delta of the Ganges; with the adjacent Countries on the East.

General Boundaries. On the West. The Hoogly and Cossimbuzar Rivers. S. The Sea. N. The Road from Moorshedabad to Jellinghy—and the Ganges, Dacca, and Tiperah Rivers. E. Aracan and Ava.

No. II. The Jungleterry District, and adjacent Provinces: comprehending the Countries situated between Moorshedabad and Bahar.

General Boundaries. W. Curruckpour Hills——Bahar Proper, and Ramgur. N. The Ganges. E. The Mauldah and Cossimbuzar Rivers. S. The Adji and Dummoodah Rivers.

No. III. South Bahar.

General Boundaries. W. Gazypour and Chunar. N. The Ganges. E. Monghir and Jungleterry. S. Palamow and Ramgur.

No. IV. North Bahar.

General Boundaries. W. Oude. N. The Napaul and Morung Hills. E. Purneah. S. The Ganges.

No. V. The Northern Provinces of Bengal: with the Bootan, Morung, and Assam Frontiers.

General Boundaries. W. North Bahar. N. Morung and Bootan. E. Assam and the Garrows. S. The Ganges—and the Road from Rajemal to Seebgunge and Dewangunge.

No. VI.

No. VI. The LOW LANDS beyond the GANGES, from the MAULDAH RIVER to SILHET.

General Boundaries. W. The Road from Moorſhedabad to Mauldah. N. The Road from Mauldah to Seebgunge and Dewangunge—and the Garrow Mountains. E. Cachar and Ava. S. The Road from Moorſhedabad to Jellinghy: and the Ganges, Dacca, and Tiperah Rivers.

No. VII. The Provinces of BENGAL, lying on the WEST of the HOOGLY RIVER: with the Maharatta Frontier.

General Boundaries. W. The Baumeen, Nagpour, and Ramgur Hills. N. The Adji, and Dummoodah Rivers. E. The Hoogly River. S. The Neelgur Hills—and the Sea.

No. VIII. The CONQUERED PROVINCES on the South of BAHAR, viz. RAMGUR, PALAMOW, and CHUTA-NAGPOUR, with their Dependencies.

General Boundaries. W. Surgoojah, Juſpour, &c. N. The Hills of Bahar and Rotas. E. Jungleterry, Pachete, and Singboom. S. The Cattack Diſtricts.

No. IX. GENERAL MAP of BENGAL and BAHAR.

No. X. GENERAL MAP of OUDE and ALLAHABAD: with Part of AGRA and DELHI.

No. XI. The COSSIMBUZAR ISLAND.

No. XII. The ENVIRONS of the City of DACCA.

No. XIII. The DOO-AB from ALLAHABAD to CALPY.

A  OF

BENGAL and BAHAR:

In VIII Parts.

GENERAL EXPLANATION.

Cities.	Post Roads.
Capitals of Provinces.	Common Roads.
Large Towns.	Passes.
Large Bazars & Cutcherries.	Fields of Battle.
Small d°.	Boundaries of Bengal & Bahar.
Villages.	Boundaries of Provinces, &c.
Forts.	

W. Harrison sc.

INDEX to the VIII *DIVISIONS* of BENGAL and BAHAR.

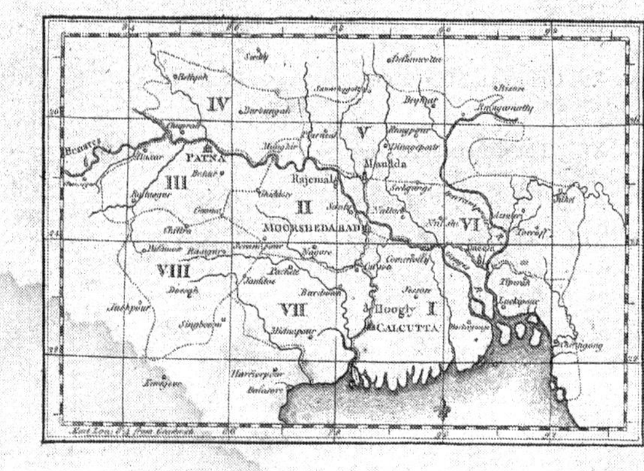

Published according to Act of Parliament by J. Rennell Nov. 20th 1779.

The
DELTA of the GANGES;
ADJACENT COUNTRIES
FRANCIS RUSSELL Esq.
J. Rennell
TIPPERAH
DACCA
JESSORE
MAHMUDSHY
BURDWAN
BAY of BENGAL

The
JUNGLETERRY
DISTRICT;
with the adjacent PROVINCES of
BIRBOOM, RAJEMAL, BOGLIPOUR &c.
Comprehending the COUNTRIES
Situated between Moorshedabad
and Bahar.
TO
Brigadier General
RICHARD SMITH,
This MAP is Inscribed
By his most obed.t Serv.t
J. Rennell.
PURNEA
BAHAR
PROPER
MAULDAH
BIRBOOM
RAMGUR
PACHETE
BURDWAN

A MAP of
NORTH BAHAR;
Inscribed to
BRIGADIER GENERAL
JOHN CAILLAUD.
J. Rennell
MUCKWANPOUR
BETTIAH
GOORACKPOUR
CHAMPARAN
MORUNG
TYROOT
PURNEAH
SARUN
HAJIPOUR
GAZYPOUR
SHAWABAD
BAHAR PROPER

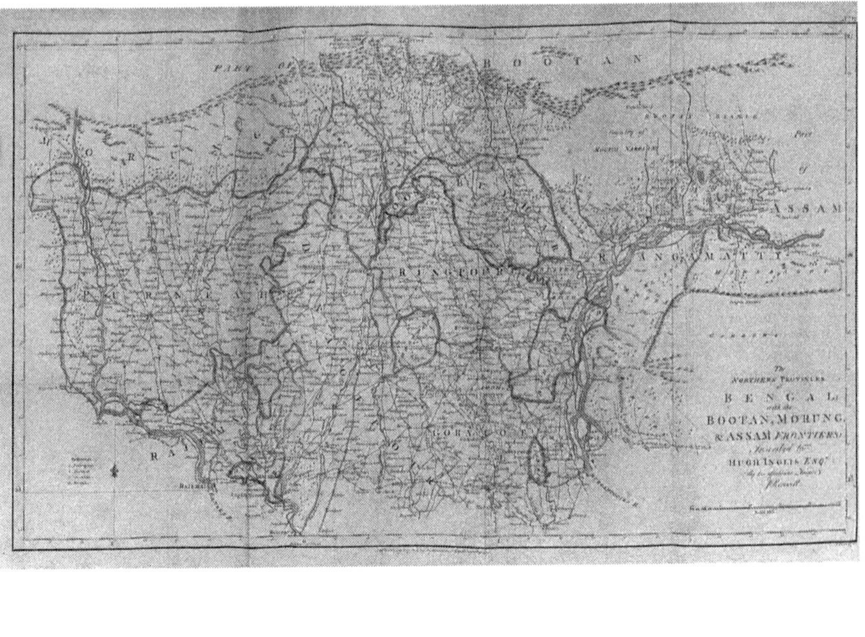
PART OF
BOOTAN
ASSAM
RANGAMATTY
RUNGPORE
PURNEAH
The
NORTHERN PROVINCES
of
BENGAL
with the
BOOTAN, MORUNG,
& ASSAM FRONTIERS
HUGH INGLIS ESQ.
J. Rennell

A MAP of
NORTH BAHAR,
BRIGADIER GENERAL
JOHN CAILLAUD,
GOORACKPOUR
CHAMPARAN
MORUNG
PURNEAH
GAZYPOUR
SHAWABAD
BAHAR PROPER

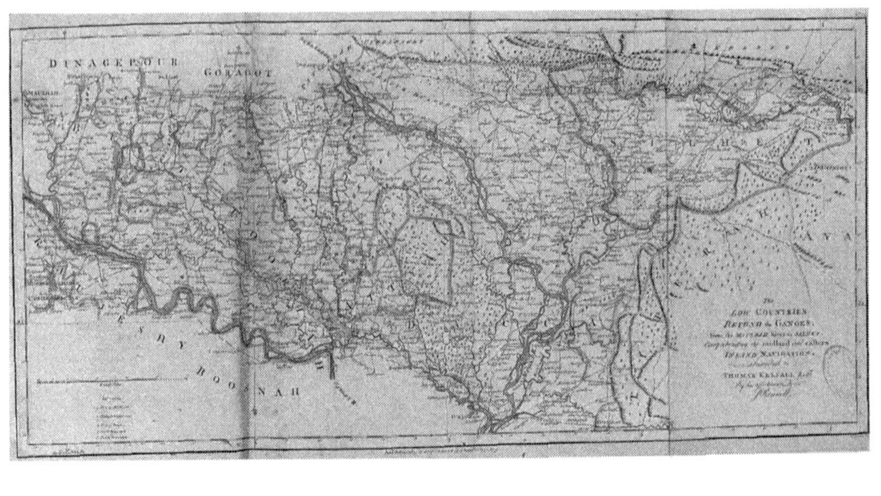
DINAGEPOUR
GORAGOT
The
LOW COUNTRIES
BEYOND the GANGES
INLAND NAVIGATION
THOMAS KELSALL Esq.

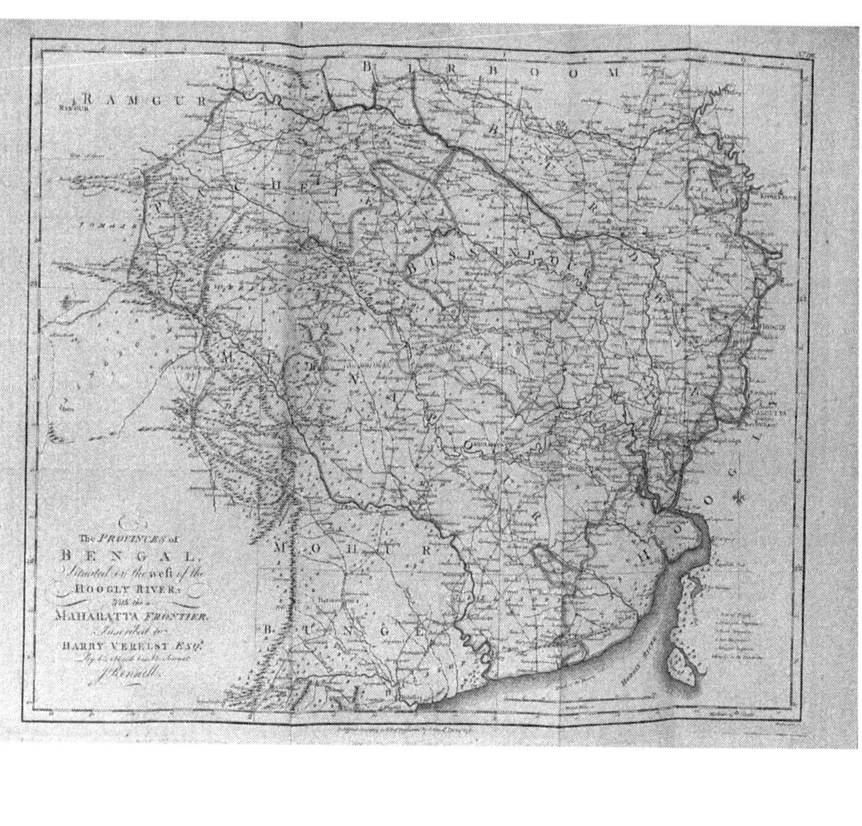
The PROVINCES of
BENGAL,
Situated on the west of the
HOOGLY RIVER:
With the
MAHARATTA FRONTIER.
Inscribed to
HARRY VERELST ESQ.
J. Rennell.
RAMGUR
BIRBOOM
BISSINPOUR

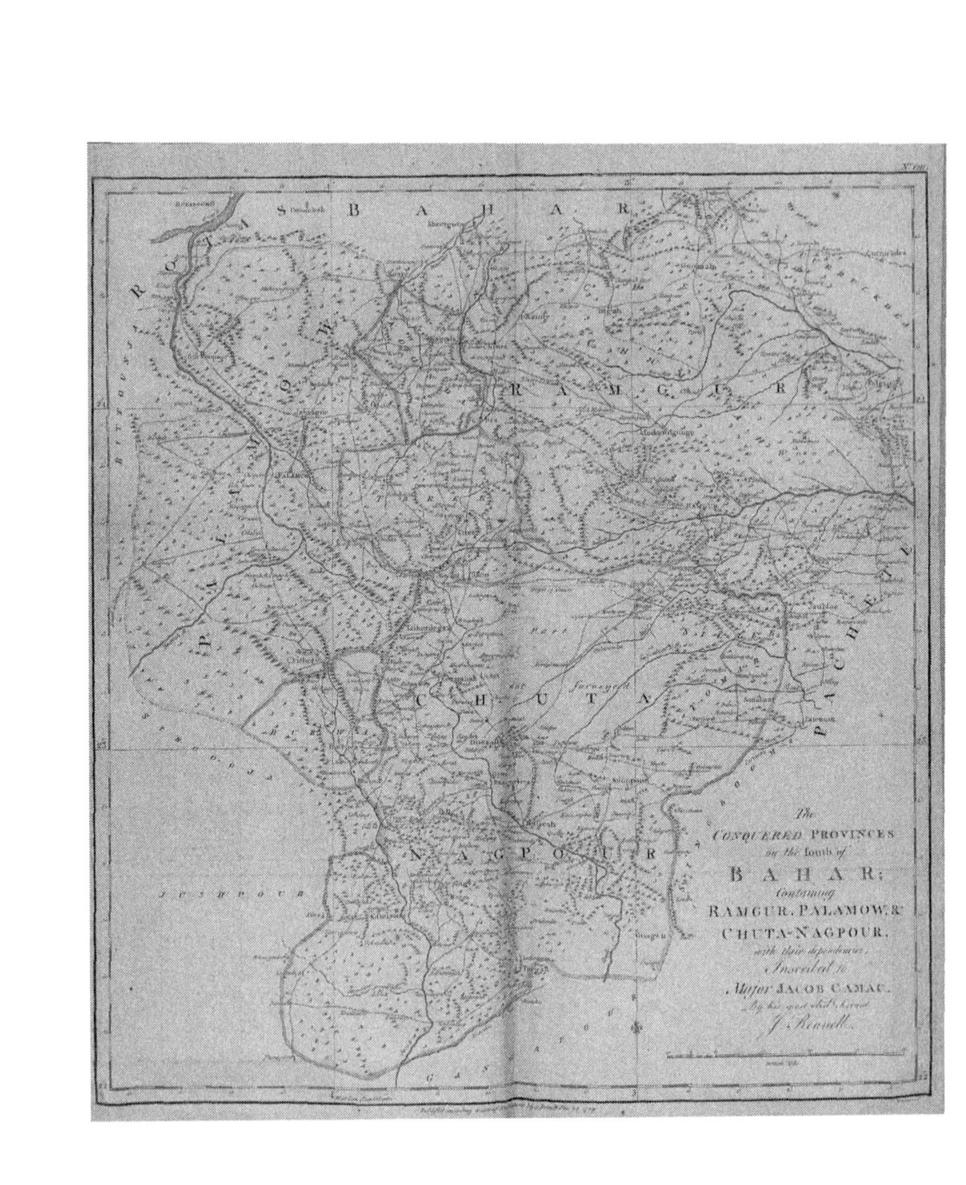
The
CONQUERED PROVINCES
on the South of
B A H A R;
containing
RAMGUR, PALAMOW, &
CHUTA-NAGPOUR,
with their dependencies;
Inscribed to
Major JACOB CAMAC,
By his most obed.t Serv.t
J. Rennell.
B A H A R
R A M G U R
C H U T A
N A G P O U R
P A C H E T
P A L A M O W
R O T A S
S
J U S H P O U R
Jaragerah

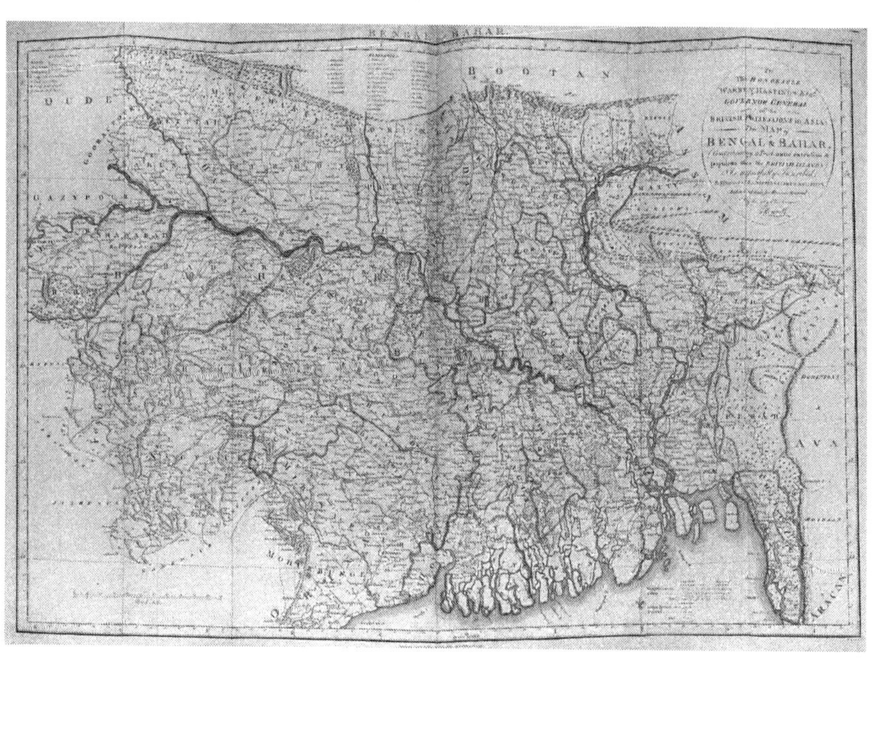
BENGAL & BAHAR.
BOOTAN
OUDE
GAZYPOUR
AVA
To
The HONORABLE
WARREN HASTINGS Esq.
GOVERNOR GENERAL
BRITISH POSSESSIONS in ASIA
The MAP of
BENGAL & BAHAR,

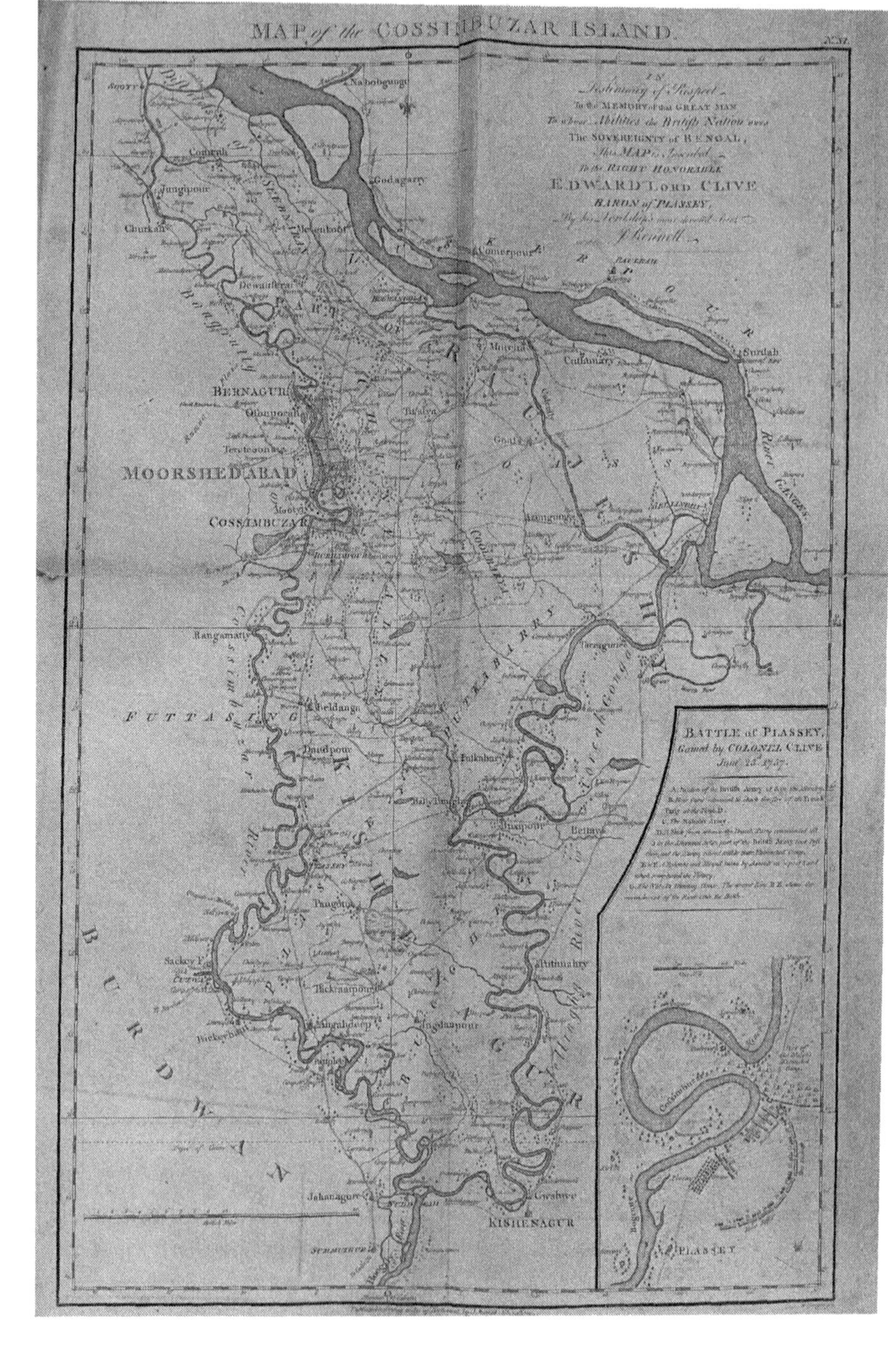
MAP of the COSSIMBUZAR ISLAND.
As a Testimony of Respect
To the MEMORY of that GREAT MAN
To whose Abilities the British Nation owes
The SOVEREIGNTY of BENGAL,
This MAP is Inscribed
To the RIGHT HONORABLE
EDWARD LORD CLIVE
BARON of PLASSEY,
By his Lordship's most devoted Servt.
J. Rennell.
MOORSHEDABAD
COSSIMBUZAR
BERNAGUR
KISHENAGUR
Nabobgunge
Godagary
Comerpour
Surdah
Cullamary
Jungipour
Churkah
Rangamatty
Beldanga
Daudpour
Pulkabary
Bettiya
Pattinahary
Jahanagure
Gwalpara
Sackey F.
Buckerbhaut
FUTTASING
BURDWAN
River Ganges
BATTLE of PLASSEY,
Gained by COLONEL CLIVE
June 23d. 1757.
PLASSEY

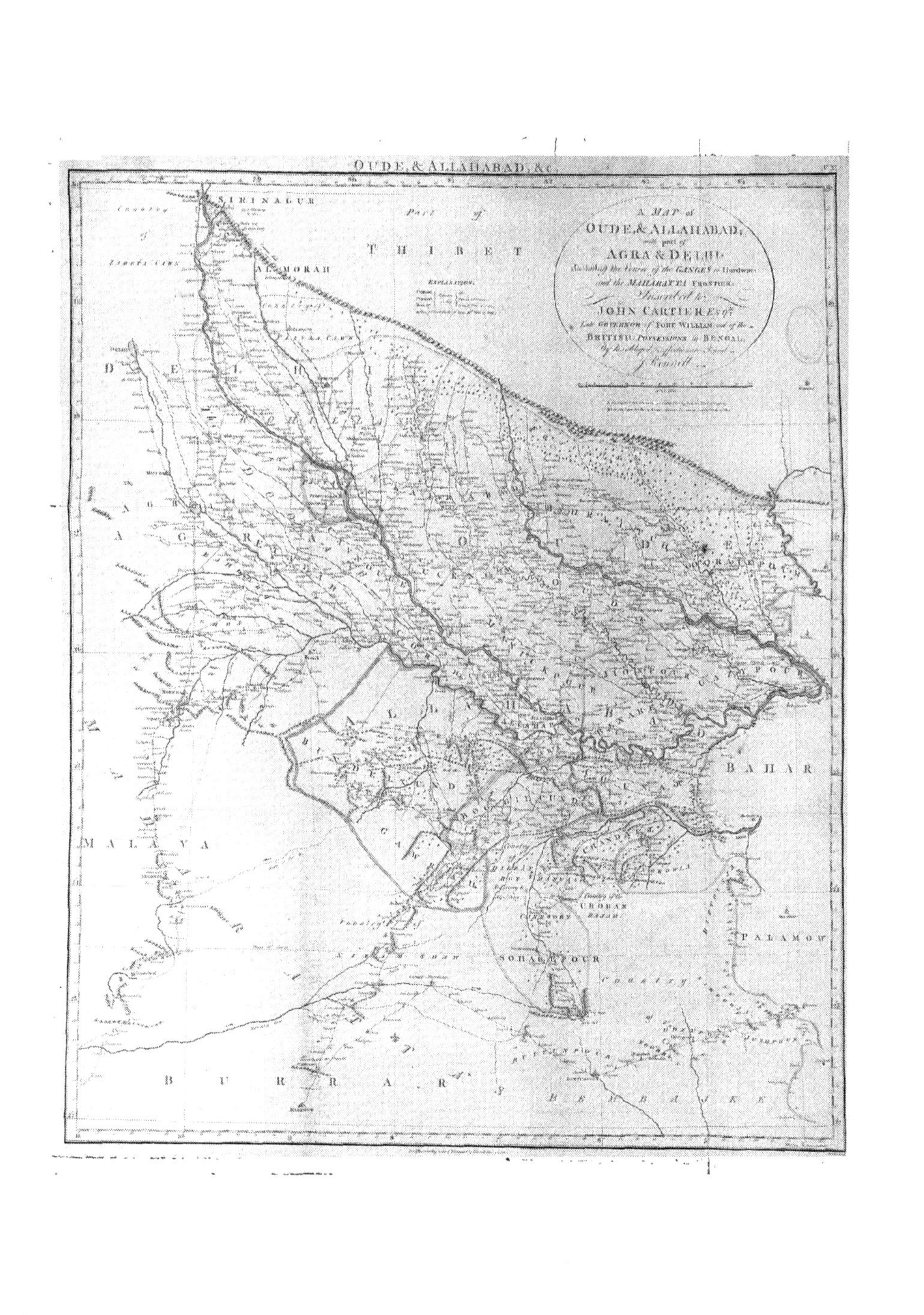
A MAP of
OUDE, & ALLAHABAD,
with part of
AGRA & DELHI:
Including the Course of the GANGES to Hurdwar:
and the MAHARATTA FRONTIER.
Inscribed to
JOHN CARTIER Esqr.
Late GOVERNOR of FORT WILLIAM and of the
BRITISH POSSESSIONS in BENGAL,
By his obliged & affectionate Friend
J. Rennell
SIRINAGUR
Part of
THIBET
ALMORAH
EXPLANATION.
DELHI
AGRA
MALAVA
OUDE
BAHAR
PALAMOW
SOHAGEPOUR
BURRARS
CROHAN RAJAH

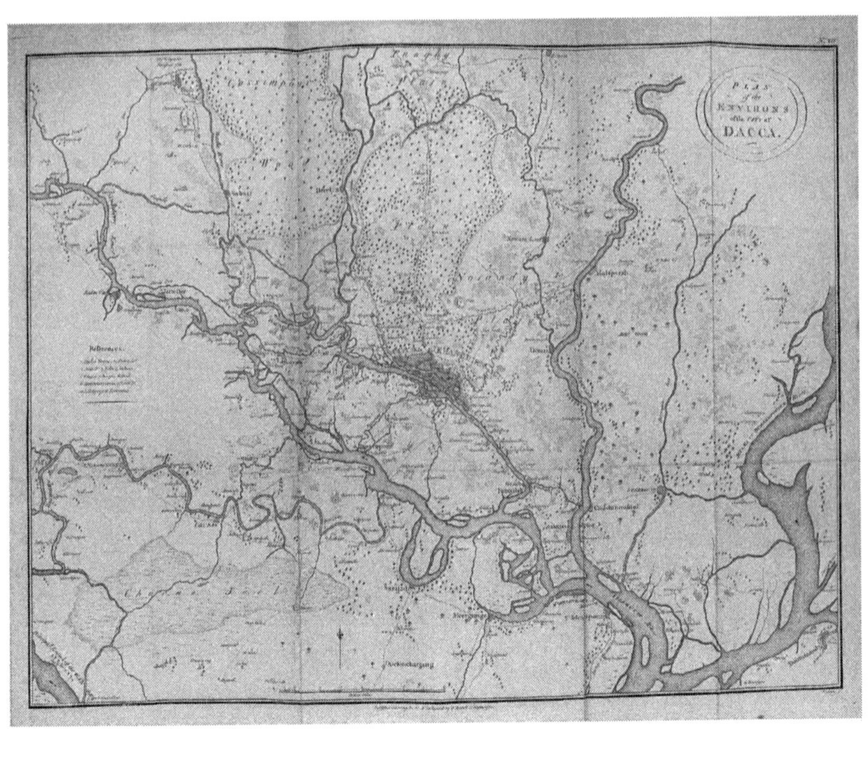
PLAN
of the
ENVIRONS
of the CITY of
DACCA.

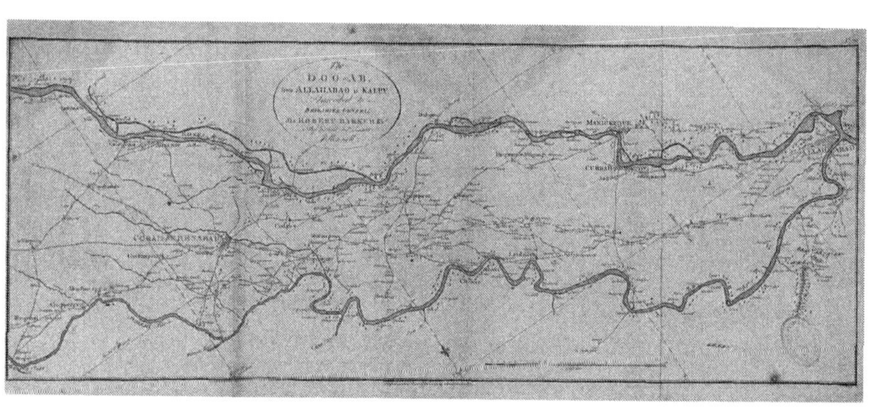
The
DOO-AB,
ALLAHABAD
KALPY

BENGAL
ATLAS
129

Milton Keynes UK
Ingram Content Group UK Ltd.
UKHW032345120224
437723UK00008B/850

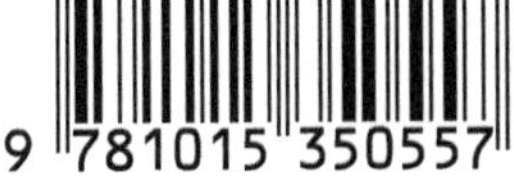